i

Baseball Statistics Helper

Efficiency Tips for the Serious Gamer

POCOL PRESS
Published in the United States of America
by Pocol Press
320 Sutton Street
Punxsutawney, PA 15767
www.pocolpress.com

By J. Thomas Hetrick

Publisher's cataloging-in-publication

Names: Hetrick, J. Thomas, author.
Title: Baseball statistics helper: efficiency tips for the serious gamer / by J. Thomas Hetrick.
Description: Punxsutawney, PA: Pocol Press, 2024.
Identifiers: LCCN: 2024930104 | ISBN: 979-8-9852820-8-5
Subjects: : LCSH Games--Anecdotes. | Board games. | BISAC GAMES & ACTIVITIES / General
Classification: LCC GV1312 .H87 2024 | DDC 794--dc23

Library of Congress Control Number: 2024930104

Illustrations compiled from numerous newspapers from the 19th century.
All are believed to be in the public domain.

Contents

This book is dedicated to the legions of fans who follow the sport of baseball, in all of its glories and disappointments. More especially, the author offers a shoutout to all those fans who cheer for historically downtrodden teams, the also rans, and the bottom feeders.

GENERAL INSTRUCTIONS

1. All *italicized* text is indicative of sample data to be inserted pertaining to your specific replay season. In order to compile all of the statistical data, it is assumed that scoresheets of each game are available.

2. This primer is developed in 14 sections;
   a. Pitching Rotations
   b. Player Injuries
   c. Team Fielding and Individual Errors
   d. Team vs. Team Matrix
   e. Standings (AL East and West)
   f. Season Narrative Analysis
   g. Team Narrative Analysis
   h. Team Batting Records
   i. Team Pitching Records
   j. League leaders
   k. Schedule
   l. Team and Individual Player Stats
   m. League Games Scores Log
   n. Unusual Events

3. Specific instructions are included for each section. This book will assist tabletop baseball gamers in keeping track of their replay seasons on a computer file, which can be constantly updated, for maximum efficiency. It is strictly "old school" and not to be used for gamers using a computer that assists in the playing, scoring, and calculating of baseball statistics.

4. Discipline is important for gamers to gamers to create accurate statistics while maximizing their tabletop baseball gaming experience. Before each replay series played, the gamer should consult this information to determine the pitcher and which players are ineligible due to injury. This information, when *updated every series*, allows for remarkable efficiency to calculate in-season and final season statistics.

5. As with the exchanging by the managers of the lineup cards and the umpires going over the ground rules prior to each game, so too should any specific league rules be discussed by the participating gamers before games are conducted.

6. Exhibition games should be played to familiarize the gamer(s) with the teams, their players, and the league.

7. Consider using a dice tower for all games, The dice tower allows for the dice to tumble in multiple directions before landing and creating a random result. Dice towers can also prevent cheating with dice. Dice towers can be purchased online.

8. To maximize the excitement level for a full replay season, consider playing only one series per day. During postseason play, consider one game per day. This allows for extended excitement and interest and more importantly, mimics baseball schedule reality.

9. All of the information contained herein can be adjusted, depending on the gamer's preference. The information and tables shown can also be created with a word processing application.

10. The information shown here is compiled in real time and at the end of the season, can be transferred into specific charts indicating final totals.

11. For template and demonstration purposes, the sample team information included corresponds to the American League of 1977, when the Toronto Blue Jays and Seattle Mariners were added to the league. Other sample teams and data is also included.

Pitching Rotations

Baltimore (1971 Orioles)	*Palmer*	*McNally*	*Cuellar*	*Dobson*	*SpJackson*	*SpLeonhard*
Boston						
Cleveland						
Detroit						
Milwaukee						
New York						
Toronto						
California						
Chicago						
Kansas City						
Minnesota						
Oakland						
Seattle						
Texas						

Pitching rotations have been part of baseball for a century and a half. This sample chart employs the teams of the American League in the 1980s with the example for the 1971 Baltimore Orioles staff. The ace of the staff (Jim Palmer) is in the first position, followed by the other starters in the rotation and spot starters. Regular rotation pitchers are indicated by last name. Spot starters have an "Sp" in front of their last name.

Before beginning a replay season, it's a good idea to fill out this chart for all of your teams. As injuries may occur along with great seasonal performances and less-than-desirable performance, these charts can be adjusted when needed. For instance, let's just say that screwball specialist Mike Cuellar developed some arm trouble in the middle of the season. Cuellar's ailment kept him off the roster for two weeks. One solution would be to employ Grant Jackson and/or Dave Leonard in the spot temporarily until Cuellar returns. Another scenario might include the replacement of another pitcher for poor performance.

At the end of the season, this chart can be removed from the final statistics as it will no longer be necessary.

Player Injuries (player + duration in games; **bold** indicates player eligible; due to roster sizes, teams cannot have more than *four* players injured at any given time and no more than *two* pitchers; parenthetical information indicates in which team game the player can return).

Baltimore	*Ripken 2 (6)*	*Murray 1 (16)*	*Lowenstein 4 (25)*			
Boston						
Cleveland						
Detroit						
Milwaukee						
New York						
Toronto						
California						
Chicago						
Kansas City						
Minnesota						
Oakland						
Seattle						
Texas						

Sore arm, muscle strain, knee contusion, sore rib, cracked rib, broken leg/arm.

Each box corresponding to a team player injury should have three data elements, the player's last name, the injury duration in games and in parentheses, the team game in which that player will return. For example, a player named Smith in injured for 10 games in his team's 23rd game of the season. The notation would then read "Smith 10 (34)" since the player was injured for a ten-game duration and will serve team games 24-33 ineligible.

Depending on team roster sizes, which vary over baseball history, the maximum amount of players injured on a team at any given time can be adjusted depending on the gamer's preference. These are displayed above in red. Replacement players can then be inserted on a team's roster for the duration of the injury.

As the replay season progresses, so too will the information in the Player Injuries table. At the end of the replay season, transfer this information onto the Team and Individual Player table in the Inj. column. Once this information has been transferred, the Player Injuries table can be deleted.

Team Fielding and Individual Errors

Team	E	Players
Baltimore	*16*	*Ripken 2 Murray 2 Bumbry 1 Palmer 1 Boddicker 1 Dauer 3 Cruz 4 Roenicke 2*
Boston		
Cleveland		
Detroit		
Milwaukee		
New York		
Toronto		
California		
Chicago		
Kansas City		
Minnesota		
Oakland		
Seattle		
Texas		

Following each series, the team and individual error totals should be placed into the table. For example, if the Baltimore ballclub made 4 errors as a team in a three-game series. The number 4 would then be played in the error column corresponding to the team. In the same row for Baltimore, the individual players who commit the errors should be documented.

This sequence should then be repeated for each team and each player committing the errors until the end of the Replay season. When the season concludes, this data should be transferred to the E (for error) column in the Team and Individual Player Stats table. Once this information has been transferred, the Team Fielding and Individual Errors table can be deleted.

Fielding percentage information based on put outs and assists are not recorded in this table.

Team Vs. Team Matrix (read row for wins and column for losses)

Team W/L	Bal	Bos	Cle	Det	Mil	NY	Tor	Cal	Chi	KC	Mn	Ok	Sea	Tx
Baltimore	■	2	2	2	3	3	1	1	2	2	2	2	1	1
Boston	1	■												
Cleveland	1		■											
Detroit	1			■										
Milwaukee	0				■									
New York	0					■								
Toronto	2						■							
California	2							■						
Chicago	1								■					
Kan. City	1									■				
Minnesota	1										■			
Oakland	1											■		
Seattle	2												■	
Texas	2													■

Following each series, the Team Vs. Team Matrix should be updated. This is accomplished by using integers inside the box corresponding to games contested. For example, if Baltimore defeats Boston by taking two out of three games in a series, the number "2" should be placed in the Baltimore row under the Boston column. For Boston, the number "1" should be placed in their row under the Baltimore column.

When displayed across or in a row, each team's entry includes wins against their respective opposition. The numbers displayed in a column indicate the amount of losses against respective rival teams.

The Team Vs. Team Matrix, along with other data, serves as a powerful indicator of each team's strengths or weaknesses.

AL East Standings

Team	W	L	Pct	GB	1-Run	2-Run	Extras	ShO	Home	Road
Baltimore	24	15	.615	--	5-2	1-3	2-1	4	16-6	8-9
Boston										
Cleveland										
Detroit										
Milwaukee										
New York										
Toronto										

AL West Standings

Team	W	L	Pct	GB	1-Run	2-Run	Extras	ShO	Home	Road
California										
Chicago										
Kansas City										
Minnesota										
Oakland										
Seattle										
Texas										

Team Standings data includes team wins (W), losses (L), percentage of wins (PCT), games behind (GB), won-lost record in one-run games (1-Run), won-lost record on 2-run games (2-Run), won-lost record in extra-inning games (Extras), shutouts by team (ShO), won-lost record at home (Home), and won-lost record in road games (Road).

The percentage (Pct) figure is calculated as a three decimal point number based on wins divided by total number of games. This number determines each team's position in the standings. The games behind calculation is based on the difference in the amount of wins and losses by the first place team when compared to all other teams. This calculation must occur twice, one for wins and one for losses. The difference in games equating to the wins total is added to the games equating to the loss total and divided by 2. This number will be the games behind (GB) number.

Gamers may decide to track other team standings data such as last ten (L10), which displays the team's record in the last ten contests. Other considerations may include runs scored (RS), runs allowed (RA), and the differential in runs scored versus runs allowed (Diff).

Season Narrative Analysis

The is a free-form narrative field with every discretion given to the gamer (author) of the information. It can include a summation of the Replay season as a whole, based on the gamer's intimate knowledge in the playing of the games (solitaire or head-to-head), the results, and the statistical data contained herein.

When writing about the Replay season, gamers might consider how the pennant race played out, key players and teams, important games, records achieved, injuries, player/manager ejections, and an overview of league play as it applies to batting or pitching dominance.

Creativity in this section is paramount.

Team Narrative Analysis

Baltimore	
Boston	
Cleveland	
Detroit	
Milwaukee	
New York	
Toronto	
California	
Chicago	
Kansas City	
Minnesota	
Oakland	
Seattle	
Texas	

The Team Narrative Analysis allows for a textual summation of each team in the Replay season.

The Team Narrative Analysis write-ups can include the same type of methodology employed in the Season Narrative Analysis. However, these narratives are team based.

The inspiration for the Team Narrative Analysis came from the initial seasons of the Sports Illustrated Major League Baseball Game, wherein a textual summation of that team's season was encapsulated in about 10-12 sentences.

As with the Season Narrative Analysis, a little creativity goes a long way here.

Team Batting Records

Batting	AB	H	2B	3B	HR	RBI	R	BA	SO	BB	SB-A	Inj.	E
Baltimore													
Boston													
Cleveland													
Detroit													
Milwaukee													
New York													
Toronto													
California													
Chicago													
Kansas City													
Minnesota													
Oakland													
Seattle													
Texas													
Totals													

Standard baseball batting statistics include at bats (AB), hits (H), doubles (2B), triples (3B), home runs (HR), runs batted in (RBI), runs scored (R), batting average (BA), strikeouts (SO), bases on balls (BB), injuries in games (Inj.), and errors (E).

All numbers in the above table are integers except for the batting average (BA). This figure is expressed as a three digit decimal number based on hits (H) divided by at bats (AB).

Other batting statistics could include slugging average (SA), on-base percentage (OBP), on-base plus slugging (OPS), and wins above replacement (WAR).

The statistics in this section can be cut-and-pasted from the Totals row in the Team and Individual Player Stats tables which exist for each team in the Replay season. It can then be sorted depending on the gamer's preference using a spreadsheet file. Normal sorting methodology might be by batting average from highest to lowest and earned run average from lowest to highest.

The Totals section here is the sum of all of the data for the entire league.

League Leaders

Batting Average		
Player	*Team*	*BA*

Earned Run Average		
Player	*Team*	*ERA*

The League Leader tables display the top ten batters and pitchers in each Replay season based on batting average and earned run average, respectively. Each row indicates the player, the team, and the average. However, any such league leader tables can be created based on the gamer's preference. For example, gamers may wish to display home run leaders or the top pitchers' win totals.

Team Pitching Records

Pitching	W	L	G	GS	CG	IP	H	SO	BB	ER	R	ERA	ShO	Inj.	E
Baltimore															
Boston															
Cleveland															
Detroit															
Milwaukee															
New York															
Toronto															
California															
Chicago															
Kansas City															
Minnesota															
Oakland															
Seattle															
Texas															
Totals															

Standard pitching statistics include games won (W), games lost (L), games pitched (G), games started (GS), complete games pitched (CG), innings pitched (IP), hits allowed (H), strikeouts (SO), bases on balls (BB), earned runs allowed (ER), runs allowed (R), earned run average (ERA), shutouts by the individual pitcher (ShO), injuries in games (Inj.), and errors committed (E).

The statistics in this section can be cut-and-pasted from the Totals row in the Team and Individual Player Stats tables which exist for each team in the Replay season. It can then be sorted depending on the gamer's preference using a spreadsheet file. Normal sorting methodology might be by earned run average from lowest to highest.

The Totals section here is the sum of all of the data for the entire league.

Schedule (italicized bold indicates series completed)

3 Games	*Bos-StL*	Cle-Det	Chi-Phil	Was-NY	
	Det-Bos	Phil-StL	NY-Cle	Was-Chi	
	Bos-Phil	Det-NY	StL-Was	Cle-Chi	Stats
	NY-Bos	Phil-Was	Chi-Det	StL-Cle	
	Bos-Was	NY-Chi	Phil-Cle	Det-StL	
	Chi-Bos	Cle-Was	StL-NY	Det-Phil	
	Bos-Cle	Chi-StL	Was-Det	NY-Phil	Stats
3 Games	StL-Bos	Det-Cle	Phil-Chi	NY-Was	
	Bos-Det	StL-Phil	Cle-NY	Chi-Was	
	Phil-Bos	NY-Det	Was-StL	Chi-Cle	Stats
	Bos-NY	Was-Phil	Det-Chi	Cle-StL	
	Was-Bos	Chi-NY	Cle-Phil	StL-Det	
	Bos-Chi	Was-Cle	NY-StL	Phil-Det	
	Cle-Bos	StL-Chi	Det-Was	Phil-NY	Stats
1 Game	Bos-StL	Cle-Det	Chi-Phil	Was-NY	
	Det-Bos	Phil-StL	NY-Cle	Was-Chi	
	Bos-Phil	Det-NY	StL-Was	Cle-Chi	Stats
	NY-Bos	Phil-Was	Chi-Det	StL-Cle	
	Bos-Was	NY-Chi	Phil-Cle	Det-StL	
	Chi-Bos	Cle-Was	StL-NY	Det-Phil	
	Bos-Cle	Chi-StL	Was-Det	NY-Phil	Stats
3 Games	Bos-StL	Cle-Det	Chi-Phil	Was-NY	
	Det-Bos	Phil-StL	NY-Cle	Was-Chi	
	Bos-Phil	Det-NY	StL-Was	Cle-Chi	Stats
	NY-Bos	Phil-Was	Chi-Det	StL-Cle	
	Bos-Was	NY-Chi	Phil-Cle	Det-StL	
	Chi-Bos	Cle-Was	StL-NY	Det-Phil	
	Bos-Cle	Chi-StL	Was-Det	NY-Phil	Stats
1 Game	StL-Bos	Det-Cle	Phil-Chi	NY-Was	
	Bos-Det	StL-Phil	Cle-NY	Chi-Was	
	Phil-Bos	NY-Det	Was-StL	Chi-Cle	Stats
	Bos-NY	Was-Phil	Det-Chi	Cle-StL	
	Was-Bos	NY-Chi	Cle-Phil	StL-Det	
	Bos-Chi	Was-Cle	NY-StL	Phil-Det	
	Cle-Bos	StL-Chi	Det-Was	Phil-NY	Stats
3 Games	StL-Bos	Det-Cle	Phil-Chi	NY-Was	
	Bos-Det	StL-Phil	Cle-NY	Chi-Was	
	Phil-Bos	NY-Det	Was-StL	Chi-Cle	Stats
	Bos-NY	Was-Phil	Det-Chi	Cle-StL	
	Was-Bos	Chi-NY	Cle-Phil	StL-Det	
	Bos-Chi	Was-Cle	NY-StL	Phil-Det	
	Cle-Bos	StL-Chi	Det-Was	Phil-NY	Stats
4 Games	Bos-StL	Cle-Det	Chi-Phil	Was-NY	
	Det-Bos	Phil-StL	NY-Cle	Was-Chi	
	Bos-Phil	Det-NY	StL-Was	Cle-Chi	

	NY-Bos	Phil-Was	Chi-Det	StL-Cle	Stats
	Bos-Was	NY-Chi	Phil-Cle	Det-StL	
	Chi-Bos	Cle-Was	StL-NY	Det-Phil	
	Bos-Cle	Chi-StL	Was-Det	NY-Phil	Stats
4 Games	StL-Bos	Det-Cle	Phil-Chi	NY-Was	
	Bos-Det	StL-Phil	Cle-NY	Chi-Was	
	Phil-Bos	NY-Det	Was-StL	Chi-Cle	
	Bos-NY	Was-Phil	Det-Chi	Cle-StL	Stats
	Was-Bos	Chi-NY	Cle-Phil	StL-Det	
	Bos-Chi	Was-Cle	NY-StL	Phil-Det	
	Cle-Bos	StL-Chi	Det-Was	Phil-NY	Stats

Creating and following a schedule is of vital importance to the gamer's experience in managing a replay season. Start with an accurate and balanced schedule for the games to be played. The schedule should be followed precisely.

The above table displays, from left to right, the amount of games to be played in a series and the participating teams on each row.

Schedules can be created from actual major league schedules which are available online and in numerous baseball-related publications. Absent an actual major league schedule, consider using an online schedule creator, which displays the schedules for generic team names. When creating your season schedule, the generic team names can be replaced with your specific league teams.

The information in this table represents a mockup of the 1908 American League Season. From 1901-1960, both the National and American Leagues featured eight teams in each league, without divisions. The winners of each league met in the World Series beginning in 1903. As such, each team played each other 22 times during the season for a total of 154 games (22 x 7). Each team contested 11 games against their rivals at their home ballpark, and 11 games away. This remarkable and classic symmetry lasted 60 years. Following 1960, with expansion and inter-league play, major league schedules have expanded to 162 games, with varying unbalanced schedules.

Compiling and calculating individual and player statistics should occur whenever the word "stats" appears in the table corresponding to that row. Statistical efficiency can further be attained by compiling and calculating following *each series* in the row that contains 'stats.' In the table indicated above, the Boston at Philadelphia series' stats can be compiled and calculated following the conclusion of that series. This compilation obviously involves statistical data generated from all of the games in the stats sequence. In other words, Boston's games would include the series' this club played in Philadelphia, hosting Detroit, and at Philadelphia for a total of nine games (bolded in red). The Philadelphia team stats could also be compiled and calculated thusly.

At the end of each series on the "stats" row, the corresponding team statistics could be compiled and calculated. This methodology ensures that all of the team statistics would be calculated and compiled at the end of the final series in the row.

Once all of the statistics are compiled and calculated, the Team Batting, League Leaders, and Team Pitching tables can be filled in. Instructions containing that task are included herein.

Team and Individual Player Stats
year Baltimore Orioles

Record:

Player	First	POS	AB	H	2B	3B	HR	RBI	R	BA	SO	BB	SB-A	Inj.	E
Totals															

Pitchers	First	W	L	G	GS	CG	IP	H	SO	BB	ER	R	ERA	SHO	Inj.	E
Totals																	

Tracking and compiling statistics is a major part of any season. Statistics should be compiled on a pre-determined schedule without fail.

Standard baseball batting statistics include at bats (AB), hits (H), doubles (2B), triples (3B), home runs (HR), runs batted in (RBI), runs scored (R), batting average (BA), strikeouts (SO), bases on balls (BB), injuries in games (Inj.), and errors (E).

All numbers in the above table are integers except for the batting average (BA). This figure is expressed as a three digit decimal number based on hits (H) divided by at bats (AB).

Other batting statistics could include slugging average (SA), on-base percentage (OBP), and wins above replacement (WAR).

Standard pitching statistics include games won (W), games lost (L), games pitched (G), games started (GS), complete games pitched (CG), innings pitched (IP), hits allowed (H), strikeouts (SO), bases on balls (BB), earned runs allowed (ER), runs allowed (R), earned run average (ERA), shutouts by the individual pitcher (ShO), injuries in games (Inj.), and errors committed (E).

Other pitching statistics could include saves (SV) and walks and hits divided by innings pitched (WHIP).

year Boston Red Sox

Record:

Player	First	POS	AB	H	2B	3B	HR	RBI	R	BA	SO	BB	SB-A	Inj.	E
Totals															

Pitchers	First	W	L	G	GS	CG	IP	H	SO	BB	ER	R	ERA	SHO	Inj.	E
Totals																

year Cleveland Indians

Record:

Player	First	POS	AB	H	2B	3B	HR	RBI	R	BA	SO	BB	SB-A	Inj.	E
Totals															

Pitchers	First	W	L	G	GS	CG	IP	H	SO	BB	ER	R	ERA	SHO	Inj.	E
Totals																

year Detroit Tigers

Record:

Player	First	POS	AB	H	2B	3B	HR	RBI	R	BA	SO	BB	SB-A	Inj.	E
Totals															

Pitchers	First	W	L	G	GS	CG	IP	H	SO	BB	ER	R	ERA	SHO	Inj.	E
Totals																

year Milwaukee Brewers

Record:

Player	First	POS	AB	H	2B	3B	HR	RBI	R	BA	SO	BB	SB-A	Inj.	E
Totals															

Pitchers	First	W	L	G	GS	CG	IP	H	SO	BB	ER	R	ERA	SHO	Inj.	E
Totals																

year New York Yankees

Record:

Player	First	POS	AB	H	2B	3B	HR	RBI	R	BA	SO	BB	SB-A	Inj.	E
Totals															

Pitchers	First	W	L	G	GS	CG	IP	H	SO	BB	ER	R	ERA	SHO	Inj.	E
Totals																

year Toronto Blue Jays

Record:

Player	First	POS	AB	H	2B	3B	HR	RBI	R	BA	SO	BB	SB-A	Inj.	E
Totals															

Pitchers	First	W	L	G	GS	CG	IP	H	SO	BB	ER	R	ERA	SHO	Inj.	E
Totals																

year California Angels

Record:

Player	First	POS	AB	H	2B	3B	HR	RBI	R	BA	SO	BB	SB-A	Inj.	E
Totals															

Pitchers	First	W	L	G	GS	CG	IP	H	SO	BB	ER	R	ERA	SHO	Inj.	E
Totals																

year Chicago White Sox

Record:

Player	First	POS	AB	H	2B	3B	HR	RBI	R	BA	SO	BB	SB-A	Inj.	E
Totals															

Pitchers	First	W	L	G	GS	CG	IP	H	SO	BB	ER	R	ERA	SHO	Inj.	E
Totals																	

year Kansas City Royals

Record:

Player	First	POS	AB	H	2B	3B	HR	RBI	R	BA	SO	BB	SB-A	Inj.	E
Totals															

Pitchers	First	W	L	G	GS	CG	IP	H	SO	BB	ER	R	ERA	SHO	Inj.	E
Totals																

year Minnesota Twins

Record:

Player	First	POS	AB	H	2B	3B	HR	RBI	R	BA	SO	BB	SB-A	Inj.	E
Totals															

Pitchers	First	W	L	G	GS	CG	IP	H	SO	BB	ER	R	ERA	SHO	Inj.	E
Totals																

year Oakland Athletics

Record:

Player	First	POS	AB	H	2B	3B	HR	RBI	R	BA	SO	BB	SB-A	Inj.	E
Totals															

Pitchers	First	W	L	G	GS	CG	IP	H	SO	BB	ER	R	ERA	SHO	Inj.	E
Totals																

year Seattle Mariners

Record:

Player	First	POS	AB	H	2B	3B	HR	RBI	R	BA	SO	BB	SB-A	Inj.	E
Totals															

Pitchers	First	W	L	G	GS	CG	IP	H	SO	BB	ER	R	ERA	SHO	Inj.	E
Totals																	

year Texas Rangers

Record:

Player	First	POS	AB	H	2B	3B	HR	RBI	R	BA	SO	BB	SB-A	Inj.	E
Totals															

Pitchers	First	W	L	G	GS	CG	IP	H	SO	BB	ER	R	ERA	SHO	Inj.	E
Totals																

League Games Scores Log

Game	Score	Winner	Loser	Winning P	Losing P	Home Runs	Extras
1	*7-6*	*Boston*	*St. Louis*	*Winter*	*Dinneen*	*Ferris, Wallace*	*10*
2	*7-2*	*Boston*	*St. Louis*	*Burchell*	*Waddell*		
3	*4-2*	*St. Louis*	*Boston*	*Powell*	*Morgan*		

Documenting a Games Scores Log for each individual game is comprised of the season game number, score, the winning team, the losing team, the winning pitcher, the losing pitcher, home runs, and whether or not the game went into extra innings and the amount of innings played.

Verifying team won/lost records, winning/losing streaks, pitcher records, home runs, and extra inning data can be accomplished by using the Search feature from a word processing application. If a gamer desires further analysis, this data can also be cut-and-pasted into a spreadsheet application's sorting features.

The League Games Scores Log also serves another important function. It can assist in reconciling statistics related to game scores, winning teams, losing teams, pitching won-lost records, home runs, and extra inning affairs in the standings.

Unusual Events

Unusual events can be documented in text form for outstanding games by players, no-hitters, rain-shortened and rainout games, triple plays, player/manager ejections, or anything else. The season game number should also be concluded. This information can then be used in the Team Narrative Analysis.

Documenting the unusual events season game number can be informed by the League Scores in the previous section.

Some examples of Unusual Events appear below in red. These events correspond to a Replay of the 1908 American League season.

New York Manager Clark Griffith ejected for arguing balls and strikes. Game 38.

Game called on account of rain in 7th inning. The Senators were leading at home at the time and under major league rules, win the contest. Game 42.

Game called on account of rain in 6th inning. The Athletics were leading on the road at the time and under major league rules, win the contest. Game 155.

No-hitter pitched by Jack Chesbro of New York. Game 52. July 12, 2022.

Suspended game, called due to massive thunderstorms. Game ended in 7-7 tie and will be completed later in the year. Game 194. Final score in 16 innings was 10-7, Chicago.

Bob Ganley of Washington grounds into a triple play. Game 195.

Washington's Walter Johnson is ejected after a foreign substance was discovered on the baseball. Suspended for 7 contests. Game 260.

Cleveland Manager Nap Lajoie ejected for arguing balls and strikes and excessive bench jockeying. Game 316.

The Bleachers
during
the shower

CONCLUSION

When all of the statistical data has been compiled and calculated, ensure that the Pitching Rotations, Injuries, Errors, and Schedule are deleted. Once this is accomplished and all instructional data in this document is also deleted, the remaining information serves as a document of the entire Replay season.

As always, the gamer has total discretion as to which information in the above document will suit their unique needs.

So, let's Play Ball!

ABOUT THE AUTHOR

Baseball historian, author, and tabletop game designer J. Thomas Hetrick culls his experience in baseball gaming by playing over 7, 000 (and counting) contests. All of them are meticulously kept on scoresheets in his home. He's the author of *Baseball Stats and Stories: Confessions of a Tabletop Simulation Gamer*, which documents fully his experiences over five decades. He's also written two other books on baseball; *MISFITS! Baseball's Worst Ever Team* and *Chris Von der Ahe and the St. Louis Browns*. The latter book earned him an invitation to the National Baseball Hall of Fame in Cooperstown, New York on two occasions.

www.ingramcontent.com/pod-product-compliance
Lightning Source LLC
Chambersburg PA
CBHW081231020426
42331CB00012B/3124

9 7 9 8 9 8 5 2 8 2 0 8 5